INTRODUCTION

Grab a pen before reading on.

This book is packed with ideas for the business owner who wants to take advantage of mobile marketing and get ahead of the competition. As you go through the book there are actions steps at the end of each chapter to help you organize and decide which initiatives you can take action on immediately.

This book is perfect if you want to better understand how smartphones and tablets are changing the way consumers interact and engage with business. All of the ideas in this book are designed to help you get fast results and they're all proven to work.

So in order for this book to be the most impactful, it will be important for you to prioritize what you want to do first. I suggest you grab that pen and paper and prioritize in this fashion.

1. We have to implement this strategy right away!

2. Interesting idea, maybe we should give it a whirl.

3. Long term need to research and make a plan.

Once you evaluate your priorities be disciplined not to do the most important, but rather do the ones that will provide the highest return. You will then discover the real value of the information in this book and begin making changes that will yield great financial returns by gaining new customers and creating more loyal customers at the same time. So look for the actions steps that make you go - that's a great idea, the ones that say I can see how that will really work, and get going with those first.

Lets get started!

"Innovation distinguishes between a leader and a follower."

- Steve Jobs

Publisher: Bill Heneghan
Cover Design: Stephen Maguire

Marketing On The Move
999 Douglas Avenue, Suite 1107
Altamonte Springs, FL 32714

ISBN-10: 0-69224-705-X (alk. paper)
ISBN-13: 978-0-69224-705-1 (alk. paper)
1. Marketing 2. Mobile Marketing

Printed in the United States of America

MARKETING ON THE MOVE

YOUR CUSTOMERS ARE MOBILE. ARE YOU?

By Bill Heneghan

CONTENTS

CHAPTER 1 – MOBILE IS THE FUTURE

How to Stay Ahead of the Curve

It may sound hard to believe, because they're so commonplace nowadays, but Apple only introduced its iPhone a few years ago, back in 2007, and its iPad in 2010. *Imagine how much has happened* since then, and how much is happening right now, as you're reading this book!

Marketing is on the Move… Are You Keeping Up?

There is simply no escaping the fact that marketing is on the move, because people are on the move. We work hard, we play hard, we want to be in touch with the people we love and the information we covet, 24-hours a day, seven days a week, 365-days a year. I mention Apple and their "i" products because they were what sparked this digital migration: a kind of freedom that allowed us to step away from our desktop computers, even our bulky laptops, and take information with us: music, text messages, voice mail, the internet and, eventually, TV programs and movies, games and access to everything from turn by turn driving directions to restaurant reservations sharing our photos, thoughts and videos on Facebook, instantly.

The call was irresistible, with a low buy-in threshold – i.e. the price of a relatively cheap smart phone – and more and more people answer it every day. How many? According to a recent report released by

ComScore called *Marketing to the Multi-Platform Majority*, "Today, more than 140 million Americans own smartphones and nearly 70 million own tablets."

What is unique about this study is that it discovered that more and more of us were merging our home and work computer use with our mobile device use. This digital migration is both significant, and historic. Says the ComScore report, "In April 2013 for the first time, more than half of digital consumers in the U.S. engaged on both computers and mobile devices." So if half of us are stepping away from our computers and going all mobile, all the time in 2013, what do you think those numbers will look like in 2014? 2015? One day, desktop computers and possibly even laptops will feel as bulky, outdated and prehistoric as VHS records do today – and twice as obsolete.

The Mobile Marketing Renaissance: *From New Kid on the Block to Owning the Block in a Few Short Years*

In just a few short years, mobile marketing has become one of the premier ways to attract new customers – and retain existing ones. What was once believed to be merely a fad has firmly established itself as a must-implement strategy for any company, from a small business enterprise to a large retailer.

At the CTIA 2013 Mobile Marketplace conference recently, a number of executives voiced their opinions about how they believe mobile marketing technologies will continue to grow in the years ahead. One thing is clear, marketers are required now to not only adapt to smaller smartphone screens and a variety of programming languages and operating systems, but also to make these technologies much more simple to use, more convenient, and more consumer-centric.

In the future, mobile marketing is likely to offer much stronger consumer engagement, drawing consumers in different ways. Over the next few years, mobile marketers expect to see companies engaging in more real-time interaction with consumers using mobile devices, further solidifying consumer relationships. There's also likely to be a lot of progress in the use of intelligence to determine how consumers use their mobile, and the use of this information to enhance convenience for users.

The use of mobile devices over the next few years is likely to grow from something that is currently restricted to marketing teams and corporations, to tools that the average consumer uses to gain access to basic services. In other words, we are well on our way towards democratization of technology, which sees more people using mobile devices to gain access to the Internet, shop for products and services, just in the way that the average American now uses the Internet to do so.

Companies that want to stay ahead with mobile marketing over the next decade must stay relevant in the features that they offer consumers, and make websites and apps as simple to use as possible. Another growing trend in the future will be the use of consumer feedback to upgrade technology. Consumer feedback will be critical to the development and successful use of mobile technology for promotions in the years ahead.

All of this is to state the obvious: mobile marketing is here, it's here to stay and it's rapidly evolving, even as we speak. But you're in the right place, at the right time, because this book is designed to give you the tools you'll need, not just to "catch up" on mobile marketing but get out in front of it and prepare for a bright, mobile future.

Why should you listen to me about mobile marketing? My name is Bill Heneghan, I'm an entrepreneur, and now an author for the last 17 years I've been in the direct marketing industry focusing on one thing - conversions, which leads to more revenue for my clients and strategic partners. I've been very fortunate to be involved in several successful ventures, and have a passion for free enterprise and small business. Having been involved in digital marketing for over 14 years the last 2 years I have studied over 2 million visitors to mobile websites of our small to medium size business clients and have put the most proven concepts into one book to help you get started effectively with a mobile marketing strategy. Although no two businesses are the same the basic principals remain solid for getting great results. Mobile has exploded since the invention of smartphones and it's not slowing down yet only 10% of small to medium sized companies are taking full advantage which is why I wrote this book. To set you at ease the purpose in this book is to give you practical applications that you can implement immediately. I grew up in a family owned restaurant so I want you to understand as I wrote this book - the idea is not to bore you with pie in the sky thoughts or visions but rather to help you see how consumers are changing the way they do things and how you can take advantage of this shift in consumerism by being in the right place at the right time.

I've never been more excited to be in business because of the opportunity that mobile marketing allows the small business owner, which is why I founded SimplyFlex | Mobile Marketing Agency to help business owners take full advantage of everything mobile has to offer. So, if you've been frustrated with trying to get new leads, a little discouraged because you aren't seeing an ROI that you where hoping for or maybe you're kicking butt and getting great results and looking for some new ideas well, this book is for you. I am genuinely looking forward to sharing the things that I've learned over the last 2 years to help you create an incredible mobile strategy. By the way, I do my best to connect with every reader if they reach out on my personal Twitter @billheneghan so I would love to hear from you - your story and what you get out of implementing some of the action steps in this book.

Startling Statistics: *Mobile Marketing by the Numbers*

If you're still on the fence about the importance of mobile marketing, let me drive the point home with a few of the statistics I keep handy for my clients. As it stands now, here is the power – and the potential – of mobile marketing today:

- **82% of adults in the United States own cell phones;**
- **91% of adults keep their mobile devices within arm's reach 24/7;**

- **90% of text messages are read within three minutes of being delivered;**
- **9 out of 10 mobile searches lead to some kind of action – over half of those lead to purchases;**
- **70% of mobiles searches lead to action within one hour... it takes one month for the same percentage of desktop computer users to catch up;**
- **74% of smart phone users use their phone to help with shopping;**
- **79% ultimately make a purchase as a result of that search;**
- **Mobile coupons receive 10 times higher redemption rates than print coupons...**

Mobile advertising is very cost-effective when you look at the returns on your investment. In fact, small businesses are likely to benefit more from mobile promotional strategies than larger corporations.

I could go on and on, but I think the points are clear, which is just this: if you're not marketing to your audience where they are – which is on their mobile devices –then you're missing out on potential sales; period. Now, with those kinds of statistics, can you afford to stay on the sidelines of mobile marketing any longer?

The Digital Migration: *Is Your Small Business Missing out on the Mobile Wave?*

Staying budget-friendly is the mantra for every small business. That's partly the reason why so many small businesses have avoided jumping on the mobile bandwagon, and have studiously avoided any mention of mobile apps, discount coupons, mobile-accessible websites and mobile advertising. There is a very definite – and possibly very expensive – misconception among small businesses that mobile advertising is only for big businesses.

Nothing could be further from the truth. The fact is that mobile advertising is very cost-effective when you look at the return on your investment, and return per dollar invested. In fact, small businesses are likely to benefit more from mobile promotional strategies than larger corporations.

Developing mobile apps can be inexpensive, and making your website mobile-accessible also does not cost a lot of money. On the other hand, it exponentially increases your outreach, and makes your business products and services much more accessible to the public and potential customers.

If you are a small business owner, who currently has a functional website, not optimized for the mobile user, then you are painfully behind the times. In fact, as many as 31% of all searches by potential customers who could be searching for your products and services, are probably not even accessing your website.

That's because your website is likely not mobile-friendly, or optimized for a smart phone or tablet. More Americans are now searching for products and services on their smartphones, than they are on their desktops and other devices. If you are not optimized for viewing on a smart phone, when people are searching for you, then you may be missing out on valuable phone calls or leads.

What every small business owner must know is that an increasing number of people now use their mobile devices exclusively to surf the Internet. The use of desktop computers and laptops has receded to the background, and a new generation of consumers – which also happens to be the section that is most likely to spend

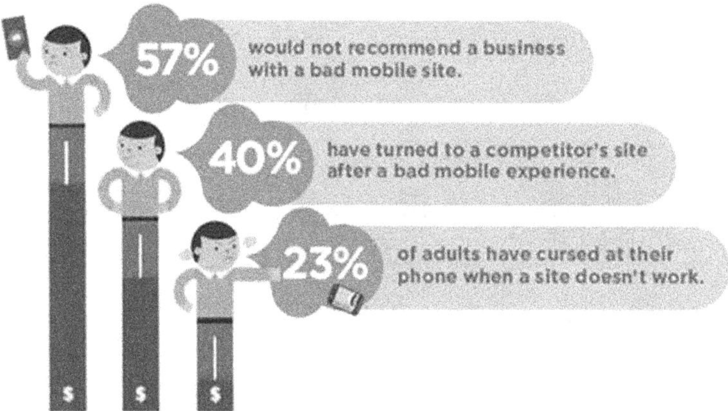

on products and services – is much more likely to access the Internet on their smart phones while on the move.

In fact, data confirms that these people are using all types of mobile devices, and not just their smartphones. They may start searching for products on their smart phones, and continue the process of researching the product and actually buying it on their tablet when they get back home. Therefore, if you are not mobile, you're missing out.

Many small business owners are not tech-friendly, and assume that anything that has to do with mobile technology has to be very tech-centered, and therefore, expensive and complicated. That simply isn't true, and in this book I'll share with you how you can easily and affordably outsource a variety of your technical needs to save time, money and those sales you're losing by not taking advantage of *Marketing on the Move*.

Conflicting Case Studies: *Success and Failure on the Mobile Frontier*

I never have to work very hard to convince my clients of the vitality and immediacy of mobile marketing because they're already living it. How many times have you been out running errands, thought about wanting to see a movie and hit up your mobile device to check the movie times?

How often have you been sitting in a restaurant when a song comes on that you just have to have. In three clicks or less, you can hit up the iTunes store, find it and download it, taking you less than a minute.

So much of our lives are linked to our mobile devices because life is on the go, and so is the future of marketing. As proof, I offer a quick case of conflicting case studies, one depicting how TO take advantage of mobile marketing, and one showing how NOT to do it:

Finnegan's Fails

Let's start with how NOT to market in the mobile age, first. Finnegan's Pub is a neighborhood eatery with a great local following but looking to expand and bring in new customers. They have a website where they feature some pictures of menu items, an outdated menu, a blog with three posts, outdated events and some contact information.

It used to work fine, but not anymore. They decide to "go mobile" and offer a single coupon event sent out to their mailing list, which contains the phone numbers and email addresses of several hundred customers.

The coupon works great; numbers go up, the listed item sells out, folks are happy and the place is jamming… but with all the same friendly faces and hardly a new customer in sight. Finnegan's has experimented with "going mobile," found it lacking and give up… for now.

Now, let's check out the same type of small, independent, local restaurant that gets mobile right:

Webster's Wins!

Webster's Café is a small local eatery featuring fresh brewed coffee ground right on the premises, a full bakery menu and a small, light breakfast, lunch and dinner menu. It's nestled on a tree-lined street in the college and research districts but would like to expand on its target audience to fill in the gaps when business gets slow.

They decide to go with mobile marketing since nearly ALL of their customers spend half their time on a phone or tablet or some sort, but want to do more than a one-off special. Working with a local company, they revamp most of their online presence, including launching across three platforms: Android, iPhone and HTML 5.

They debut a new website featuring daily coffee blends, drink specials, seasonal recipes, spoken word and musical performance listings, GPS directions, a blog and an image gallery featuring the "art" several baristas draw on the customer's coffee cups.

The website revamp and debut are in sync with the creation of a mobile app customers can download to access all of the above web features on their smart phone or tablet. Yes, there are frequent coupons, specials or "deals," but they are in line with Webster's overall philosophy of mobile marketing: go with the customers are and give them what they're looking for. As a result, they received an almost immediate 26% in social media connections and a 34% increase as a result of push notifications.

Mobile Marketing Isn't Just an Event, It's a Habit

As you can see, mobile marketing – true mobile marketing – is about more than just blitzing a coupon out to your mailing list and letting the rest work itself out.

Mobile marketing isn't an event; it's a habit. What's more, it's a mindset of facing the future, feeling a little anxious about how much – and how quickly – marketing, promotions and advertising are changing and, rather than burying your head in the sand and ignoring it, facing the challenge with a new and modern attitude.

That's what this book offers: a new philosophy, based on years of experience in an industry where daily changes make every experience priceless.

It would be impossible to teach you every mobile marketing method in this book, because by the time it was published they would likely have changed or become obsolete.

Instead, this book will help you understand what your customer needs, and how to give it to them: quickly, immediately, where they are: on their integrated and accessible mobile devices. You will understand how they think, and learn to deliver what they need where they are – and right on time.

Just as importantly, it will teach you *why* mobile is an important part of your marketing:

Reasons Why Your Business Should Go Mobile

If you are a small business owner and do not have a mobile presence yet, you are missing out on major sales potential. Here are just six reasons why your business should be smart phone-accessible.

1. **Mobile is now the number one way to access the Internet, and smartphones have left desktop computers far behind in this regard.** Obviously, you can't take your computer everywhere you go, but you can carry a smart phone in your wallet, your handbag or in your pocket. People definitely find it easier to access the web while they are on the go, and that includes not just surfing the web, but also shopping for products and services. For the overwhelming majority of Americans, the smart phone is always within reach.

2. **Smart phone use is not just high, but also expanding rapidly.** Approximately 35% of smartphone users recently admitted that they plan to increase their Internet usage via their smart phone going ahead. What that means is that there

are likely to be more smart phone users in the next year, than there are right now. The demand for smartphones clearly hasn't peaked. What is even more important for you to understand is that most phone users in the United States now have smart phones.

3. **IF A CHUNK OF YOUR SMALL BUSINESS CUSTOMER BASE IS LOCAL, THEN YOU SIMPLY CANNOT AFFORD NOT TO HAVE A MOBILE PRESENCE.** Statistics find that approximately 94% of people, who own a smart phone, are looking for local information, products or services on their mobile. Even better, 70% of these local searches end with a conversion – being a phone call or a visit to your location.

4. **WITH A MOBILE MARKETING STRATEGY IN PLACE, YOU ARE ABLE TO TARGET CUSTOMERS 24/7 LITERALLY, WITHOUT HAVING TO WORRY ABOUT PRIME TIME.** No longer do you have to worry about targeting customers in the evening when they are relaxing in front of the TV. With a mobile campaign in place, you could be in touch with customers all the time, informing them about new deals and products, boosting conversions and promoting sales.

5. **A MAJORITY OF THE POTENTIAL CUSTOMERS, WHO VISIT YOUR MOBILE WEBSITE, ARE LOOKING TO BUY.** These are people who are ready to make a purchase decision immediately. In other words, converting these customers simply does not involve the kind of trouble and effort that other types of marketing campaigns do. As many as three quarters of mobile users are ready to buy right now. If your business is not mobile-accessible, you're missing out on a huge opportunity to boost sales, and beat competitors.

6. **THIS BRINGS US TO THE LAST AND ONE OF THE MOST IMPORTANT REASONS WHY YOU SHOULD BE INVESTING IN MOBILE MARKETING.** It's very likely that most of your competitors are not mobile-friendly. Small business owners have only recently begun to wake up to the potential that mobile marketing offers, and are investing in SMS campaigns, app development, and other types of campaigns that help boost revenues. Less than 15% of business websites are optimized for smartphones and tablets. If you want a marketing strategy that leaves the competition far behind, then going mobile is the way forward!

The fact is that most people now use or access the Internet on their smart phones, and if a user finds that your website is not easily and conveniently accessible on his mobile, he simply will move on to your competitor.

Four Guiding Insights for Engaging Through Mobile

In addition to social media monitoring, we want to move into engaging your customers in a very real, very interactive, very mobile way. To begin our discussion, let me just offer the following four insights on how mobile marketing can help you engage: more often, in a more target manner, to more people, than you've ever imagined before:

Insight #1: *Start with a Mobile Web presence*

The cornerstone of any mobile strategy should be a Web presence, whether it is a single landing page designed specifically for smaller, more mobile screens or a fully integrated mobile site. Many companies are content to just "shrink" their current website down to mobile size but a truly mobile web presence is required for optimum effect.

Even if a marketer is experimenting with sending SMS messages, these messages should include a link to a page where mobile users can learn more and integrate in a way that flows fluidly for use on a wide variety of mobile size monitors.

Here are four types of mobile Web presence:

1. **Responsive designed websites.** Companies have built mobile websites which offer nearly the same features as their traditional websites, but which are adapted to a handheld format.

2. **Plug-in-based mobile sites.** Similar to the first category, blogs and websites based on WordPress, Drupal or similar open-source platforms can use free plug-ins which format such sites for mobile audiences so that they can be enjoyed with most of the same features.

3. **Mobile landing pages.** As the name suggests, these single-page entities can be created quickly to add a mobile-Web presence to a marketing campaign. While having less functionality than either items 1.) or 2.), above, they are still

preferable to simply ignoring the mobile market and forcing them to try and visit your clunky, not-mobile-marketing-ready website.

4. **Dedicated mobile sites.** These sites are standalone, multi-page entities, not merely mobile versions of a traditional website. They have their own designs and strategies to meet the needs of mobile visitors and are principally what I'll be talking about as we move forward through this book.

INSIGHT #2: *Consider All Your Mobile Options*

Mobile networks and devices provide a range of ways to reach an audience -- such as text, voice and email. When your team is considering how best to incorporate mobile into its marketing, consider all the major possibilities:

- **Short Message Service (SMS):** SMS is capable of sending minimal, text-based messages to your audience, which can include links to call a phone number or visit a website.

- **Multimedia Message Service (MMS):** MMS is similar to SMS technology, but can also send content such as images, video and audio files such as ringtones.

- **Voice:** Mobile phones have click-to-call functionality that enables audiences to reach you directly, or to click to request a call from your team.

- **Web:** Similar to traditional Web browsing, the mobile Web is continually adding pages of content designed for easy access from handheld devices.

- **Proximity marketing:** Smartphones with GPS and similar technologies are capable of broadcasting locations. Some marketers are taking the opportunity to deliver ads to mobile users in specific locations, such as when they're near brick-and-mortar stores.

- **Applications:** Computer programs specially designed for smartphones are widely available and have their own marketplaces. Some marketers have directly integrated campaigns into their audiences' phones by designing and offering a branded mobile app.

- **Content:** Branded content, such as including ringtones, images, videos and ebooks, are just a few of the many different types of digital information marketers can provide in a mobile format.

- **Email:** As any business professional with a Blackberry will tell you, email is a mobile channel. People frequently receive and send digital letters through handheld devices.

INSIGHT #3: *Mobile Doesn't Stand Alone*

Mobile marketing does not succeed as an isolated channel. Instead, it works best when integrated with other marketing, web and social media channels and tactics to form a cross-platform strategy.

For instance:

- **Combining SMS or barcode calls-to-action in traditional advertising;**
- **Mobile apps that integrate with television shows;**
- **Mobile coupons for in-store sales…**

Mobile promotions should also be integrated with other channels. For example, mobile content should be promoted on your website -- e.g. if you're advertising a free whitepaper download and it's available in a mobile format, mention this in your website ads to create synergy and exponentially multiply calls to action. More than just a marketing tool, mobile is the great connector for all other media channels.

INSIGHT #4: *Mobile Requires a Well-Planned Strategy*

It can be tempting to quickly test proximity marketing or a mobile website just to see what happens, but an ill-conceived marketing campaign with little to no planning – and even less strategy – can actually do more harm than good. Before "floating" a test case just to "see what might happen," here are some specific areas for you and your team to consider:

Overall marketing strategy

Mobile devices are extremely personal. Owners carry them everywhere, and during all stages of the buying process. Given mobile's "constantly-connected" attributes, you must understand the impact of making mobile information available throughout your entire marketing strategy, not just as an isolated (see above) or one-off situation.

To include mobile in your overall marketing strategy, your team should know:

- **Which specific goals you want to achieve;**
- **Who will be doing what to further those goals;**
- **How the tactic will help achieve those goals;**
- **What possible negative impact it could have…**

Usage cases for your audience

Determine the ways in which your audience would, or already does, interact with your company on mobile devices. Don't just Q & A in an academic sense.

Instead, put yourself in their shoes -- how could they use a smart phone to learn more about you? What would they do, realistically, to learn more, click here, go there and possibly commit to making a purchase?

By checking your website's analytics you may find mobile visitors are already accessing your site. I've seen anywhere from 30% to 50% of existing websites getting hit by mobile phones today, and if you're not prepared to take advantage of those visitors yet, using mobile as part of your overall marketing strategy is the first step in the right direction.

Parting Words

I predict, by 2015, if not sooner, mobile will be the **primary way consumers access the Internet**.

A mobile marketing strategy is critical for every business, and so on the following pages I will not only lay out the case for mobile marketing and your business future, but provide a simple, straightforward strategy every business can follow, of any size, in any industry...

CHAPTER 1:
ACTION STEPS

Here are three simple Action Steps to help make Chapter 1 more immediate:

Step 1: *Run a Spot Check*

Before you go any further, stop what you're doing and check to see how your website looks on your smart phone right now. It should be mobile friendly, accessible and optimized for viewing and user experience. Here's how to check:

1. Is it pinch and pull? (In other words, do you have to use your fingers to read the content?)

2. Is there a clear call to action for click to call?

3. Is there an easy way to click for directions so a customer can quickly touch a button to get door to door directions?

4. If you have Google analytics see how many mobile visitors you are actually getting?

Answering these four simple questions will immediately tell how "mobile friendly" your current website is – or if it needs improvement.

OR TEXT MOTM TO 407-641-1345

As you can see, if you scanned or opted in to receive the sample how interactive print can become as a result of an integrated strategy.

CHAPTER 2 – MOBILE IS SOCIAL

THE MOBILE CONNECTION

"NOTHING STOPS AN ORGANIZATION FASTER THAN PEOPLE WHO BELIEVE THAT THE WAY YOU WORKED YESTERDAY IS THE BEST WAY TO WORK TOMORROW."

– *JON MADONNA*

Social Media, Mobile Marketing and You: *A Crash Course*

Before we go much further in our discussion of the power of mobile marketing, I want to push the "pause" button for a moment and address how social media is changing the way consumers consume, and marketers market.

We all know the sites: Facebook and Twitter being the "top dogs" but plenty of worthy rivals to include Instagram, Klout, Pinterest, Tumblr, etc. These social media titans drive much of our modern conversation online, and fill much of our idle – and not-so-idle – time.

Social media quickly became a way for people to interact with each other in a fun, safe, non-threatening way. Suddenly we could all post family photos, quotes, tributes, jokes, updates, multimedia, whatever our hearts desire. Companies quickly learned that social media was also a great place to market their wares, and commerce quickly became a part of that online fabric.

But what many companies fail to understand are the basic reasons why people use social media, and I can tell you from experience that it's not to shop.

Engagement Matters

Social media matters because it gave immediacy to our online communications. Before social media we still had email, forums and other ways to post, but nothing compared to the instant gratification – and communication – of posting a photo on Facebook or "live" tweeting an emotion, thought or news item.

Businesses call such items "posts," but consumers think of them as "engagement." As other words, each share, each like, each follow, each tweet or retweet isn't just a post to them, it's a matter of communicating, no different than picking up the phone, sending an email or leaning over the backyard fence and chatting with their neighbor.

But too many companies see all posts as merely that: posts. Snippets of electronic communication disguised – if at all – as engagement:

- **A potato chip company asking your favorite chip flavor;**
- **A movie studio wishing you Happy President's Day;**
- **An author offering their eBook for 99-cents over the holiday season;**

- **A magazine offering a discounted rate for Mother's Day;**
- **A recording artist offering a sample of their new hit record;**
- **Etc.**

While such posts may increase page views or brand recognition, they do little to truly engage the consumer into making an actual purchase. Or, if they do move a consumer to purchase, they do little to retain brand loyalty past the point of purchase.

Why? Because the consumer was never fully engaged. Frankly, because the consumer was always treated like a consumer – not a true connection, a valued friend or even part of the company's "family".

This lack of engagement is why social media is so full of unintended clutter, making it hard for folks to be social or engaged in any real or meaningful way. Many companies treat social media much as they do traditional advertising: as one-way "ads" simply published in a new medium.

But one dimensional advertisements that require little, if any, input from the social media user and thus stay with them about as long as any other random factoid that blitzes across their subconscious while browsing through their Facebook wall or Twitter feed.

So, what is true engagement? True engagement is what happens when you connect with a consumer in such a way that it makes an impact. It's offering not just news, but news they can use. It's offering not just coupons, but the right coupons at the right time for what they're looking for. What's more, increasingly, true engagement isn't just about what they're looking for – it's where they're looking for it.

That's why mobile marketing is becoming so commonplace. As more and more consumers use their smart phones the way they used to use their desktop computers, then their laptops and later their tablets, they are increasingly looking to their phones as a mobile office, home entertainment suite, news source, catalog, GPS and movie theater all in one.

Their phones and other mobile devices now account for all the old, traditional places we used to market: magazines, TV screens, movie theaters, newspapers, even postcards and letter campaigns as they get their emails, texts and alerts all in the same place.

As we wind down our journey through the power of social media marketing and, in particular, mastering the art of mobile applications for your business, I want to stress one of its most important features: **mobile is transparent**.

Never before in history has so much information about the general public and, in particular, your target market been so readily available, and all voluntarily.

Everything we do online – but more importantly, everything your customers do online – is there for public eyes to see. Our Twitter feeds, our Facebook profiles, our websites or blogs, the comments we make on blogs or our favorite websites, it's all there just waiting to be discovered.

Consumers are taking to social media to air their complaints about everything from bad fast food drive thru service to ineffective dishwasher detergents to faulty running shoes.

The Power of Social Media Monitoring

As little as a decade ago, there was very little social media to monitor. Perhaps the mother of all social media sites, MySpace, was founded in 2004. Facebook, its natural offshoot, was only founded in 2004 and Twitter started two years later, in 2006.

In that time social media has grown from a casual entertainment and "niche" market to the preeminent form of networking, marketing, communication and, increasingly, customer service. It is how we communicate, all of us, on a daily basis.

As social media became more popular, so did companies who dabbled in what we now call "social media monitoring," or observing online chatter to take the pulse of today's modern consumer. What also began as a fringe science is now a full-fledged industry, and even that is steadily evolving.

According to MonitoringSocialMedia.com (www.monitoring-social-media.com), "In a recent Gleanster report, 73% of 'Top Performers' (the top 25% of those surveyed) identified Customer Service as a top reason to invest in social media monitoring – only 1% behind PR."

Like pretty much everything else these days, smart, savvy and sophisticated consumers are taking to social media to air their complaints about everything from bad fast food drive thru service to ineffective dishwasher detergents to faulty running shoes.

As a result, modern brands employ teams of individuals to monitor their online "chatter," scouring Twitter feeds and Facebook timelines, Tumblr images and Vine videos to see what people are saying about them, their products, their employees and, ultimately, their service.

The beauty of social media is also a double edged sword. While it's great when our customers talk us up online, thanks to the ultimate transparency of social media, it can be damaging – even disastrous – when they talk us down:

How Mobile Marketing Can Help You Protect Your Company's Online Reputation Better

There are immeasurable benefits to be gained in having an online presence for your company. In fact, a company that cannot be found online these days simply does not exist. However, there can be a flipside to online presence. The Internet, especially with the use of mobile devices, makes it very simple and very easy for people to vent about your company's products or services, if they are not satisfied with these. Online reputation management, therefore, is a critical service that every small business owner should be familiar with.

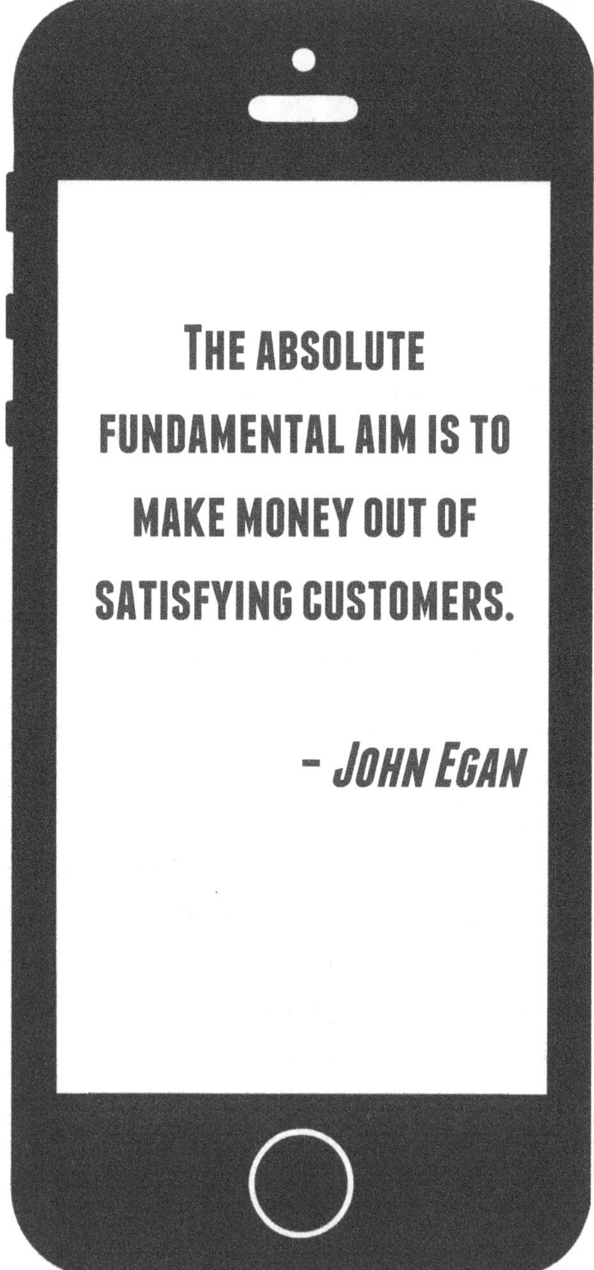
THE ABSOLUTE
FUNDAMENTAL AIM IS TO
MAKE MONEY OUT OF
SATISFYING CUSTOMERS.
- JOHN EGAN

You don't have to be a Fortune 500 company to have a reputation to protect. If you are a small business owner that is focused on the local market, you need reputation protection just as much as a larger company. Negative feedback on Facebook, Twitter, Yelp, Trip Advisor or the dozens of other outlets can quickly damage your carefully built reputation.

Say, for instance, you have an e-commerce customer who is not satisfied with the service provided by your company, after a delayed shipment. The rule of thumb, as far as Internet marketing goes, is that customers are much more likely to post negative reviews of a product or service, than a glowing review. If a customer is not satisfied with your response after the delayed shipment, he won't wait long to post a negative comment on your Facebook page. That comment is likely to be viewed by thousands of fans on your Facebook page, instantly creating a negative impression about your company in the minds of potential customers.

Fortunately, you can access these comments quickly and read them as soon as they are posted on your Facebook page on your smartphone. It is important to instantly respond to such feedback. The best way to respond is to apologize for the inconvenience caused, and reassure the customer that the problem will be fixed as quickly as possible.

Not every negative comment posted online has to spell disaster for your business. In fact, a smart business owner will take a negative comment as a springboard to build up a more satisfactory relationship with the customer. In the above example, a smart business owner will not just respond to the customer's comment, but also get to work on fixing the problem for the customer as quickly as possible, so that the customer is satisfied. Most customers, who find that their problem is fixed quickly after posting a negative review, are indeed satisfied, and their negative views of the company immediately fade away. They will now be in the mood to again do more business with you, and this depends entirely on how you have resolved the matter. A smart business owner will keep the lines of communication open to a formerly disgruntled customer, to ensure that he returns to the business.

Online reputation management therefore, does not necessarily have to mean simply limiting damage to your company's online reputation from negative feedback, but also advising companies on using this feedback to improve products and services, and enhance customer experience. Naturally, it is important for businesses to understand all of the social conversations that are taking place.

5 Reasons You'll Want to Monitor Social Media

Monitoring social media will help you boost, as well as defend, your online brand. Here are the **Top Five Reasons You'll Want to Monitor Social Media**:

1. **It's Where Your Customers Are:** Not listening to online chatter is like ignoring customer complaints, surveys and any other type of traditional feedback. We have to get away from thinking that technology is intimidating, overwhelming or too complicated. It is simple another way of listening to your customers – and who doesn't want to do that?

2. **They're Talking Directly To You:** Nowadays, everyone and their brother knows that social media is transparent. So when someone trashes – or praises – your product, service, employee(s) or brand, it's because they secretly or not-so-secretly want you to know about it. Not listening does you both a big disservice.

3. **It Will Help You Keep Your Online Brand Integrity Intact:** Nowadays, we all live in fear of

online transparency. Most of us have experienced that single customer who, for whatever reason, makes it his or her life's mission to destroy our online reputation. Often this is a disgruntled customer who could have been appeased if only someone had been monitoring his online chatter and reached out to head him off at the pass. This is easier when you have someone – or several someones – monitoring social media and preventing such PR disasters before they ever occur.

4. **Stay Abreast of Trends:** In addition to assisting you in your ongoing efforts to master the art of customer service, monitoring online chatter can help you keep abreast of – and possibly even predict – future trends.

5. **Give Customers More of What They Really Want:** Many of us live in a corporate bubble, setting up task forces and think groups to determine what our customers want… from the inside out. When we monitor social media, we take advantage of its ultimate transparency to see beyond our own ivory towers to discover what our customers really want… in their very own words. Think of it as an online focus group, at our

service, 24-hours a day, seven days a week, 365-days a year!

As you can see, there are multiple big- and small-picture reasons why social media monitoring is important. What's nice is that mobile marketing can be a part of the solution, not the problem.

CHAPTER 2:

ACTION STEPS

Here are three simple Action Steps to help make Chapter 1 more immediate:

Step 1: *Check your Social Media Profiles*

Where are you putting your time into social media – and is it the best fit for your company? Facebook and Twitter might be the "big dogs" in this hunt, with billions of users, but are they really the best platforms for your creative team? Your product? Your service? Your industry? Your personality?

Google+, YouTube, LinkedIn, Pinterest, Tumblr, Instagram, Foursquare, TripAdvisor and Yelp are all social media outlets where consumers are having social conversations that you should know about. It's better to invest in some quality time getting to know which ones are the right ones for you than to spend all time on one that might not be, or too much time spreading yourself too thin over all of them.

Also, make sure your business is listed properly across the most popular social media sites and mobile apps. We have created FREE Report to show you how your business is showing up on over 45+ apps and social sites - like Google+, Yahoo, Bing, Facebook, Foursquare, Yelp and many more. If there are errors or in consistencies in your listings, fix them immediately – this will also help with your SEO as a side benefit.

Access free report here:
simplyflex.com/motmreport

Step 2: *Monitor Your Online Reviews*

In today's consumer driven market, people rely heavily on what others say. They want quick insight in a sea of information, and increasingly user reviews on sites like Amazon, TripAdvisor and Yelp are guiding consumer choices.

When you see negative reviews about your own organization, product, service or personnel, ignore them at your own peril. Instead, take the time to reply and show others who visit these sites that you are engaged and care what people say.

We all understand that folks can leave bad reviews for a variety of reasons and informed consumers know to take them with a grain of salt. However, when a company takes the time to address them – calmly, politely, reasonably but regularly – it says a lot about that company's integrity and can often undo the negative review that prompted the response in the first place.

For social media to be truly impactful, it needs to be a discipline across social media outlets based on your internal marketing strategy.

Step 3: *Don't Just Post, Engage*

Invest in real engagement, not just posting on Facebook or Twitter accounts to "fill" your various social media pages. For example, Facebook recently made changes to its algorithm to make sure your content is what they consider "valuable". If it isn't, chances are it is not going to be seen by your target audience and, therefore, your posts are just that: simply "posts" with very little impact.

Step 4: *Outsource to the Right People*

The way consumers are engaging with all types of business is changing forever. It isn't going away and, in fact, mobile marketing is something you need to embrace.

It is important to connect with a firm that truly understands social media and put a plan in place that will help you create great content. I've found that it is much more affordable to partner with a firm than try to hire a social media "expert" full time.

For a small business, creating a basic strategy should start around $499 month and ensure that you are getting a quality return on your social media efforts – and investment.

Step 5: *Check Your Stats*

For a quick check on your social media – most have a Facebook page – check your "insights" to measure your reach on each post. For instance, if it is only 50 to 100 per post, you want to definitely connect with a social media firm that can help you with a greater impact

CHAPTER 3 – MOBILE IS NOW

The Power of Instant Communication

An iPod, a phone, an Internet mobile communicator... these are NOT three separate devices! And we are calling it iPhone! Today Apple is going to reinvent the phone. And here it is.

- *Steve Jobs*

Mobile is Now.

The title of this chapter can be taken in one of two ways. For starters, you can take it to mean that mobile is *now*, meaning it's in the present tense, it's happening as we speak and is only going to be more commonplace with each passing day.

Secondly, and just as importantly, mobile is **NOW**! In other words, it signifies instant communication, not just between friends and family, but immediate access to a wide spectrum of products, services, brands and organizations, such as:

- **Instant movie tickets;**
- **Instant dinner reservations;**
- **Instant music downloads;**
- **Instant videos;**
- **Instant airline tickets;**
- **Instant rental cars;**
- **Instant information;**
- **Instant access;**
- **Instant purchasing power…**

The fact is, studies show that mobile search has increased 400% since 2012. No, that's not a typo. 400%. The only other thing that increases that much in less than a year might be… a plague!

Plague or no, it's becoming increasingly clear that mobile marketing is impacting our lives in every way, connecting our physical and digital worlds in ways that might have seemed like so much science fiction as little as five years ago.

Could you imagine, back then, seeing a phone in nearly every single hand as you:

- **Dine in a restaurant?**
- **Wait for a plane?**
- **Sit in a movie theater?**
- **Lounge in a hotel lobby?**
- **Walk into a conference room?**
- **Etc.?**

Today, we have a device that is entertaining us, informing us and changing us in ways we never imagined.

Phones are ever present, ever evolving and all encompassing. Today, literally in the palm of our hands, we have a device that is entertaining us, informing us and changing us in new and exciting ways, never imagined before.

Yes, watching a movie on our phones while we wait in the doctor's office is nice. Listening to music on our phones while we walk on our lunch break is great. Recording our thoughts or tapping out notes or texting multiple people while talking on speaker phone is revolutionary. But perhaps what's most interesting about the mobile revolution is that it allows us to immediately take action:

The Figures Are In: *The Power of Mobile NOW*

To cement the link between mobile technology and immediacy of marketing, allow me to share just a few more numbers from our friends at Anchor Communications.

While reading these, pay careful attention to the growing disconnect between email and text usage and the rise of SMS, or Short Message Service (mobile) usage:

- The average person looks at their phone 150 times per day!
- SMS produces 6-8 times MORE engagement than email!
- Only 20% of all emails are read.
- Only 12% of Facebook posts are read.
- Only 29% of tweets are read.
- A whopping 98% of SMS messages are read!
- 70% of people in the United States have stated that they would like to receive offers from brands on their mobile devices.

- 68% of people in the United Kingdom have stated that they would like to receive offers from brands on their mobile devices.
- The average response time for an email (business to community) is 2 days.
- The average response time for a text message is only 90 seconds!
- Email only experiences a 4.2% click-through rate.
- SMS experiences a whopping 19% click-through rate!
- SMS coupons are 10 TIMES more likely to be redeemed!

The takeaway from this column of startling numbers, for me anyway, is that people are beginning to look at email, and even social media promo like tweets, as somehow less vital than the SMS messages they receive on their phone.

So, why all these numbers? Are they important? To me, they're critical in divining the future, a glimpse into trends that aren't just distant but happening right now, as we speak. Recognizing the value in such statistics can prove to be the difference between keeping up – or even falling behind – and staying ahead of the competition.

One aspect of mobile marketing that I find particularly interesting, and that cropped up so often in the above polling of statistics, is the issue of proximity marketing:

People are beginning to look at emails, and even social media promo like tweets, as less vital than the SMS messages they receive on their phone.

What Is Proximity Marketing and What Can It Do for Your Business?

As using mobile devices to target customers becomes easier and more cost-efficient for companies, a new cutting-edge strategy called "proximity marketing" is redefining the way organizations interact with customers whenever they are near a business.

Proximity marketing relies primarily on the use of mobile devices. Simply put, this kind of marketing uses Wi-Fi technologies as well as Bluetooth™ to transmit messages of specific businesses to a cluster of mobile devices within a certain radius.

For instance, you might find a company like a retail business sending out targeted messages about an upcoming sale or discounts to customers within a geographical radius on their tablet computers, mobile phones and other mobile devices. Customers might see that, figure they're in the area anyway, so… why not stop by?

Proximity marketing is already being used very effectively by retail companies, and the industry which is already a highly competitive one, is finding that being able to target customers based on geographic location is very cost-efficient and effective.

Take for example, a major retailer which is trying to bump up sales in a particular mall where sales of its products are traditionally low. In such cases, an effective strategy can be to run a discount campaign to drive up sales. Typically, a company like this might use print space to advertise its campaign, but this would be a much more expensive tactic.

The company could also use social media like its Facebook or Twitter pages to advertise such campaigns, in addition to – or in lieu of – a traditional print campaign. However, the problem with these "limited reach" campaigns is that they only reach those people who've already signed up for those notifications, or who are already following the company's Facebook or Twitter feed. Other people will not receive these notifications.

However, with proximity marketing, the company can install physical routers in the store, effectively transmitting the message that they intend to send out to customers within a specific geographical range. Suddenly, any person in a certain geographical territory who has access to a smart phone or tablet computer can receive the message informing them about the discount sale, directly on their smart phone screen via Bluetooth or Wi-Fi. Now the store's reach extends beyond its social media and opt-in followers to attract a potentially new and loyal fan base of local customers.

This is a strategy that is not just effective for retailers, but also for restaurants and bars. Anyone in the restaurant or bar industry knows that heavy profits are to be had by designating happy hours, or special menus for a fixed time only.

Say, for example, you run a bar that wants to push sales by designating a weekday happy hour. You could use the above mentioned technique, and target the local office crowd that is on its way back home, tired and in need of a drink.

Your restaurant or bar may not be tops on their minds at the moment. That is, until they get an "alert" that reminds them of how close you are, the deal you're running and, look… it's just around the corner on the way home from work!

Proximity marketing is still too new to say for certain whether it can guarantee success in almost every field. However, the fact is that in a crowded marketing arena, any company that tries out something new and innovative to reach out to consumers, is already ahead of the game, particularly when we're looking at the kinds of growth, opportunity and numbers depicted by the statistics I've shared in this chapter.

Long Term Trends Are Emerging: *Mobile + Social + Local*

The rise of social media has been well documented, as Facebook surpassed 1 billion users in the blink of an eye. But in this rapidly evolving world of modern marketing, to stay in one place is never advised. Already social media is beginning to merge and morph and splinter as new sites like Tumblr and Snapchat threaten to topple those that were present at the beginning, like Facebook and LinkedIn.

More and more, mobile marketing continues to grow as consumers move from a desktop computing environment to computing on the go. They enjoy the freedom, and immediacy, of mobile marketing as opposed to being tied to a desktop, den or home office computer, and this "on the go" technology is providing increased profits for those who can take advantage of emerging trends.

So, what are some other mobile trends I see emerging as we take a walk through tour of opportunities on the emerging horizon? Here are a few that I see having a long-range impact on the future of NOW:

Click-To-Call

This may seem counterintuitive to all we've been told about modern marketing but, according to Google, consumers would rather call a business than log-on to their website when they are researching a product on their mobile device.

If you think about it, this actually makes sense to anyone that has tried to navigate a complicated website on a tiny mobile screen. It is often much easier to simply click on a provided link to a 1-800 number and call. Click-to-call functionality is much easier to use for consumers.

Google also reports that when the marketers added the click-to-call feature in their Ad Words advertisements, they experienced an increase of 63 percent in the ratio of click-throughs. With that level of success, expect to see more click-to-call screens when you research products and services on mobile devices.

Better Mobile Email Marketing

Digital marketing firm Knotice reported that mobile drives 36 percent of total e-mail opens. This is an important figure because e-mail is one of the major activities that mobile users do on their devices.

Smart marketers are optimizing their e-mails to make sure that they appear properly on mobile platforms, paying closer attention to things like ease of graphics, simpler/tighter formatting, print color and font sizes, etc. Optimizing e-mail for mobile will continue to be a major focus for aggressive marketing firms in the months to come.

Parting Words: *The Computer in Your Pocket*

The message that these statistics and, hopefully, this book are sending is clear: If your small business marketing strategy is not currently mobile-centric, it needs to be. There's enough evidence to indicate that consumer access of the Internet through smart phones and tablets is not just high, but is steadily increasing.

Statistics also indicate that more than 50% of consumers have made the switch from browsing the Internet on their desktop computers, to accessing the web through their tablet computers and smartphones.

There are a number of reasons for this, not least being the fact that smartphones are now not just cell phones which are used to talk and text with, but little miniature computers that allow users to do everything from browse and surf, to shop online.

This leads us to our next topic – Mobile is SEARCH:

CHAPTER 3:

ACTION STEPS

The mobile consumer is a now your potential customer. Increasingly, smartphone users want information faster, and easier to access. Consumer phones are always on, always connected, and always with them. Take a quick look at the way customers stroll your retail store, website, or event and make it easy for them to engage with you as well as incentivize them.

Here are two things you can do right away to easily begin building a list of customers with mobile:

Step 1: *Look for Ways to Connect with Smartphone Users through Mobile.*

Consumers are willing to give up a little bit of information for rewards and offers. Building your list of potential customers is one of the aspects that helps yield high returns on your mobile investment. Build a database so you can remarket strategically to your existing customers and attract new ones in a more targeted, personal and profitable way.

Step 2: *Start a Mobile Rewards Program*

There are many SMS (short service message) platforms to choose from, including Protexting and ExactTarget. Personally, I use SimplyFlex Mobile. Incentivize customers to join your list by offering a discount or incentive them in exchange for their phone number and email. I have outlined a Rewards Program Blueprint step by step if you visit simplyflex.com/chapter3.

CHAPTER 4 – MOBILE IS SEARCH

How Customers Use Their Smart Phones

"If you want something new, you have to stop doing something old"

— *Peter F. Drucker*

We talk about life before the Internet, but today the web is no longer the future, it's the present. It's how we live; right now, today. We want our news immediately, online, in digital format. And we don't want to wait for it; we search for what we want, in every department: news, movies, music, restaurant menus, the works. And the more we utilize the search figure, the more we understand that the world truly is at our fingertips.

More, faster, new, now. We want our music live and streaming, and to be able to organize it any way we wish, across a variety of platforms. We want our movies on demand, instantly, pause and play and fast forward and rewind. We want our books fast and cheap and digital, and what's more, we want all this across multiple platforms.

Welcome to the cloud:

Cloud Technology: *The Future is Here*

Cloud technology allows us to take entire libraries of information, both personal and professional, anywhere we go. Thousands of our favorites songs, all our work files and documents, all the movies our kids love, every digital picture we've ever taken, all of it is immediately available everywhere we go.

Sure, it's on our desktop computer at home, and the laptop we bring with us to meetings, but it's also on our tablets and, increasingly, our phones. We have grown so accustomed to "tablet technology," the ability to access everything we want in increasingly smaller, thinner and lighter "tablets" that it's almost seen as restrictive when we sit at our desktop computer or even open up a light, portable laptop.

Mobile marketing isn't just a fad, trend or newfangled way for publicists to toss around buzz words like SMS. It's evidence of a massive, global and permanent shift in consumer behavior.

Times are not going backward. Video rental stores aren't going to be opening back up around the block anytime soon. Music stores aren't going to see a huge resurgence. Even technology once considered "new" and "modern," like DVDs and Blu-Rays, are now seen as analog, anachronistic and soon-to-be-extinct.

Think about it: who needs to store hundreds of big, bulky DVD cases, CDs and books when you can store thousands of your favorite movies, albums and books on a cloud and access them from anywhere?

Who needs to stare at a print newspaper, magazine or billboard ad when you can see it on your mobile screen, and actually bring it into the store with you to use the coupon you've just been sent because you're within five miles of your favorite fabric, frame, running shoe or other retail business?

Nearly every day there is some technological advance, some idea, some technique or revolutionary new program that allows us to do something digitally – on the go – that we used to have to sit somewhere, plugged in, and perform on a single device.

But don't take it from me. Recent, cutting edge and targeted research supports these bold claims as well. Case in point: According to SweetStrategies.com, here are some fairly mind blowing statistics to help you understand the smart phone consumer and remember almost everyone has a smart phone these days:

- **81% of them browse the Internet on their phone;**
- **77% use a search engine on their phone;**
- **48% of them watch video on their phone;**
- **39% of them use their smart phone in the bathroom;**
- **33% of them use their smart phone while they are watching TV;**

- **22% of them use their smart phone while they are reading the newspaper;**
- **79% of them use their smart phone to help them with their shopping;**
- **74% of them make a purchase based on a smart phone search;**
- **95% of them look for local information on their phones;**
- **88% of these take action within the same day.**

What I'd like you to notice about these statistics, and don't get me wrong, every one is a game changer, but what's particularly interesting to me are the last two points:

• 95% of them look for local information on their phones

• 88% of these take action within the same day.

This is the power of search. Whether it's a product, or a service, a plant or a sneaker or a bottle of wine or a massage or a specialty chocolate, you have something your local customers need. And thanks to the way we do things these days, when we need something, we search for it.

What do we want? A running shoe, a palm tree, a couples massage, a personalized coffee mug, sea salt and caramel fudge. When do we want it? Now! So we search for it, using a simple equation:

What we want to obtain + where we are = result

This is where you come in. You have what local people want, and you are where they are, so you should be in the top ten results they get when they key in something like:

- **Movie memorabilia + Hollywood, CA**
- **Organic produce + Seattle, WA**
- **Running shoes + Las Vegas, NV**
- **Movies times + Denver, CO**
- **Disney souvenirs + Orlando, FL**

- **Etc.**

Are you? Are you even close? Have you considered how your marketing efforts reflect both SEO content and mobile search technology? If not, you're just a few steps away from being able to not just catch up to mobile marketing proficiency, but master it:

Why Small Businesses Should Prioritize Mobile Strategies

There's no denying the fact that these days, a marketing strategy that doesn't include mobile campaigns is doomed for failure. However, the stakes have been upped since QR codes and apps first made their appearance. In other words, now it's no longer sufficient for a company to merely have a functioning website that's mobile accessible, and a halfway – decent app.

For companies to truly leverage mobile, they must make mobile a top priority. That involves enthusiastically embracing all forms of mobile campaigns including SMS marketing, app development, geo-specific digitized coupons and other tools.

There are several reasons why you need to focus on mobile first as part of your marketing strategy. One of the bigger reasons is market advantage. It is very likely that you are one of the first businesses in your market constantly targeting customers 24/7, being in touch with them, interacting with a potentially new customer base and so on. You don't have to be a marketing guru to know that being first matters.

Many small businesses are experimenting with mobile marketing solutions. There's no doubt that doing so helps these companies stay up-to-date with the latest marketing technologies, and boosts innovation levels at these companies. Mobile still remains a fairly new technology, and therefore, requires a lot of thinking-outside- the- box. That simply encourages more innovative thinking at your company, and you're much more likely to come up with new ways of getting in touch with customers, improving your products and services, and boosting sales.

Mobile is a much more targeted and direct way of reaching out to customers compared to print advertising, and therefore, you are more likely to focus on the important things that ultimately boost sales in your business. In other words, mobile cuts out all the unnecessary fluff, helping you narrow down your focus in promoting your business, services or products.

It's completely safe to prioritize your mobile strategy because these technologies are fairly new, and any mistakes are likely to be overlooked or forgiven. That is not so with other types of promotional campaigns or tools that have been around for several years now, and are fairly well established. Mistakes in these types of campaigns are not likely to be overlooked as easily.

For example, the effects of a bungled print advertising campaign can reverberate for years, damaging your company and image significantly in the process. However, when it comes to mobile, things are still fairly up in the air, and any promotional experiments that are not so successful will not have a strong negative effect on your company. That definitely encourages more outrageous and innovative customer-targeting strategies!

Parting Words: *Making the Most of Mobile With an Optimized Website*

If your marketing strategy hasn't yet made place for a mobile connection, then it is high time that you got on board. The first thing to do is get a mobile-friendly website that is accessible on a smart phone. A regular website simply will not do, especially when it comes to mobile search.

A mobile-friendly website will be accessible on a smart phone and tablet computer, and will be easy to use. If your website isn't a mobile-optimized, a consumer who checks out your website on your tablet computer and finds it annoyingly difficult to access, will simply move away to a competitor, or to another company, that does have a mobile-friendly website.

If you already have a mobile-website but haven't really given it much thought, now is the time to optimize your website, to make it more accessible to a growing population of consumers that shops online.

If you already have a fairly decent mobile-accessible website, it is time to take things further, and make it easier for consumers to shop on your website. There are a number of mobile-friendly payment gateways that make it easy and safe for consumers to shop for products using their smart phone. Developing an app should also be on your list.

Consumers now increasingly use apps to keep in touch with a company's product and services, and also to look for specific products and product information when they are inside a store. Launching an integrated marketing campaign using social media, mobile, and with local intent will separate you from the competition.

CHAPTER 4:

ACTION STEPS

Astonishing as it may sound, in the very near future the primary way we will be accessing the Internet will be through smartphones and tablets. Here are three FREE things you can do to maximize your search results:

Step 1: *Google+*

Visit www.google.com/business. Make sure you've claimed your Google+ page. Many people do not even realize they have one. If you have already claimed your page, make sure it is filled out fully with your most accurate business details, hours, information on your business and, most importantly, become as active on Google+ as you are on other social media outlets. Google+ is free and will impact local search in a very positive way.

Step 2: *Google Maps*

Make sure your business is listed accurately on Google Maps. Although there is a process to this – and Google will need to verify your business location – it is well worth the time and effort to be listed accurately.

Step 3: *Believe in Video*

Finally, build a YouTube channel for your business and post videos regularly. What kind? You might want to consider posting any of the following, depending on your industry:

- **A "how to" video;**
- **A making of video;**
- **A cooking video (if you're a restaurant, etc.);**
- **A presentation video;**
- **A virtual tour;**
- **A description of your services;**

- **Grinning customer testimonials;**
- **Etc.**

These videos don't need to be a major production. Consumers are more interested in the content than Hollywood production values. You don't need an expensive camera to make this effective; today's smartphone video technology is more than sufficient.

Special Note:

The reason these three things are important is that Google owns all three outlets, yet less than 10% of businesses are taking advantage of these outlets to maximize search results. That's why it is important to have your mobile site buttoned up before taking these simple steps.

CHAPTER 5 - MOBILE IS INTERACTIVE

Join the Social Media Conversation

We've all done it: received a coupon via email or, increasingly, text message and stood at the cash register, phone out, so the cashier can scan it. It's a great way to engage customers and combine mobile marketing, but increasingly using this strategy alone is becoming the equivalent of simply sending a flyer in the mail.

But what if you could make it more interactive and engaging for your consumers? What if, for instance, you could send a "mystery" coupon that might produce more than the average "10% off" coupon presented at the register? But the customer could only find out in the store!

Or…

- **What if you could send consumers an automated reminder of their appointment, say, at your service center, dentist's office or beauty/nail salon? And, in addition to simply confirming or canceling, you could – if they decide to cancel – provide half a dozen alternate times to ensure they lock in a day and date right then?**
- **What if you could combine a print ad such as in your local newspaper or "Penny Pincher" magazine with the customer's phone by having them "view" several missing pieces from the ad**

only visible through the camera feature on your phone?

- **What if you could ask customers to vote on your next product name, packaging or promotion in exchange for a particular savings, discount or reward?**

These are just some of the ways that mobile marketing is becoming increasingly interactive, fusing lighting fast technology with almost everywhere wireless internet and the active imaginations of creative marketers – and CEOs like yourself – all over the country.

One I saw recently that I really liked was for the Shazam music and TV company app. When the familiar "S" shaped Shazam icon appeared on the screen – I saw this in a local movie theater – you could text a certain phrase (for instance, "shzm") to a certain number (i.e. 1-800-SHZAM) and be instantly entered for a chance to win a Carnival Cruise.

As you can see, the technology is already there and waiting. All it takes is a commitment to create fun, fresh and marketable ideas that both capture the public's imagination and compel them to act:

Interactivity: *The Future of Mobile Marketing*

What does the future of mobile marketing look like? In a word: **interactivity**. Above all, mobile marketing in the future is likely to offer much stronger consumer engagement, drawing consumers in different ways. Over the next few years, mobile marketers expect to see companies engaging in more real-time interaction with consumers using mobile devices, further solidifying consumer relationships.

There's also likely to be a lot of progress in the use of intelligence to determine how consumers use their mobile, and the use of this information to enhance convenience for users.

The use of mobile devices over the next few years is likely to grow from something that is currently restricted to marketing teams and corporations, to tools that the average consumer uses to gain access to basic services.

In other words, we are well on our way towards democratization of technology which sees more people using mobile devices to gain access to the Internet, shop for products and services, just in the way that the average American now uses the Internet to do so.

Companies that want to stay ahead with mobile marketing over the next decade must stay relevant in the features that they offer consumers, and make websites and apps as simple to use as possible.

Another growing trend in the future will be the use of consumer feedback to upgrade technology. Consumer feedback will be critical to the development and successful use of mobile technology for promotions in the years ahead.

Wal-Mart Investing in More Effective Mobile Marketing Strategies

The world's largest retailer is sharpening its mobile marketing focus to increase customer convenience, helping them save not only time but also money while shopping.

It's part of a new, interactive feature Wal-Mart recently unveiled at the *CTIA 2013: Mobile Marketing* conference. The company has been at the forefront of mobile marketing in the retail sector, and has over the past few years, actually magnified its mobile marketing efforts.

The point at Wal-Mart has been to develop mobile tools that are useful to the average customer who walks into a Wal-Mart store. The goal is not just to help them find the products that they need more easily, but also to help them find products that are affordable.

Wal-Mart estimates that more than 50% of its consumers walk into a store with smart phones. The company also believes, as I do, that customers who use mobile phone apps are much more active in using these applications in the store, compared to those who do not use apps.

The company says that almost 30% of its online Wal-Mart.com web traffic is driven by smartphone users. In fact, according to the Wal-Mart experience, smartphone customers are some of the most loyal customers. That is a huge advantage in the retail industry, where competition is fierce and customers have a wide variety of options to choose from.

The company has been experimenting with new and innovative ways to enhance the mobile shopping experience for customers. It has developed a shopping list feature that is highly customer-centric, and boosts the shopping experience for customers.

For instance, customers can use this tool to check the price of a particular product, and determine which particular aisle in the store the product is located in. The tool automatically calculates the customer's running basket total, as they add new products to the cart.

Not only that, the company is also brainstorming ways that customer shopping can be further enhanced,

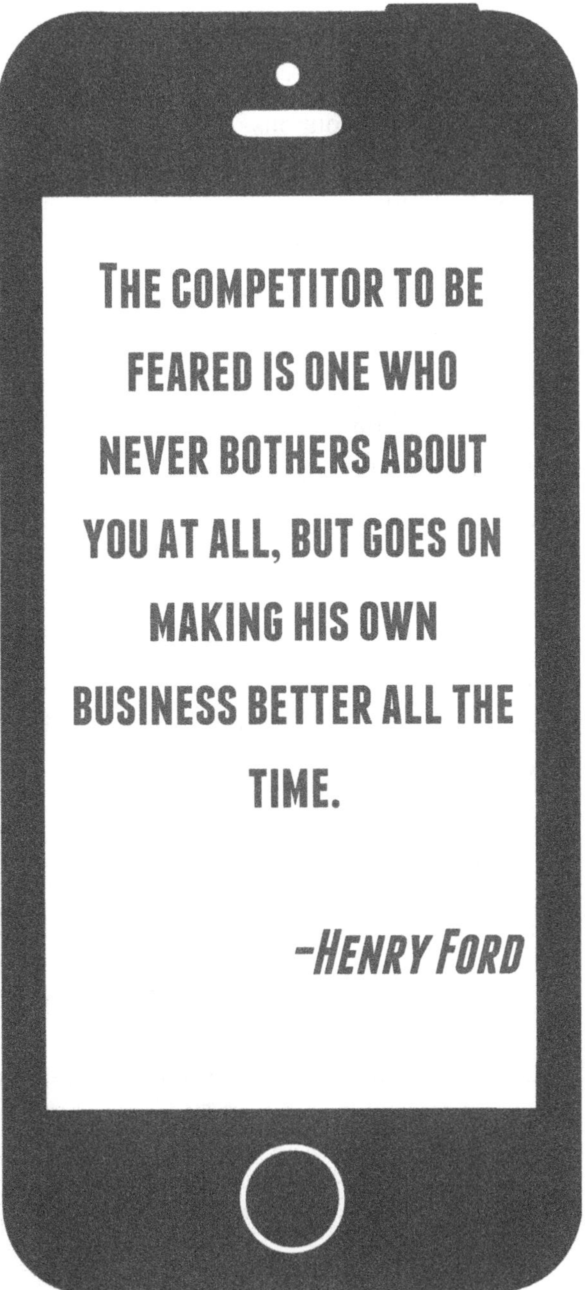
THE COMPETITOR TO BE FEARED IS ONE WHO NEVER BOTHERS ABOUT YOU AT ALL, BUT GOES ON MAKING HIS OWN BUSINESS BETTER ALL THE TIME.
-HENRY FORD

so that the tool actually recommends products to customers when they walk into the store, and displays digital and manufactured coupons.

The Wal-Mart mobile experience is one that many companies can attest to. By 2016, e-commerce sales could touch as much as $345 billion. That works out to approximately 10% of all total retail sales. Of this, smart phone-based sales or mCommerce will touch approximately $27 billion. Sales made through smartphones using mobile apps are likely to account for approximately 8% of all e-commerce sales. Companies that ignore these statistics do so at their own peril.

Case in point: seeing a retail monster like Wal-Mart implement mobile marketing strategies positions them perfectly for retail distribution outlets for online buyers anywhere.

Interactivity: *There's an App for That*

For mobile marketing, becoming more "Interactive" means more than just sending a text message and expecting a response. True interactivity comes when you create real, ongoing relationships with consumers who actively want to participate in your mobile marketing interactions.

Fortunately, mobile phone users are becoming more and more familiar, in fact, they are demanding applications they can download straight to their phone for more convenience, portability and "playability" right at their fingertips.

While not every company can offer an app that plays thousands of songs like Spotify or Pandora, or offer a national gaming sensation like Candy Crush or Plants Vs. Zombies, technology – and the demand for it – is such that there are a variety of applications – or "apps" – to make nearly every company more interactive.

For instance, there are a variety of companies who specialize in creating downloadable apps for local restaurants who want to increase interactivity among their customers, and find new customers.

These apps are simple, attractive, functional and provide numerous benefits for both the restaurant and the customer. For instance:

- **The app provides driving directions, which prompts the user to input a location or address, offering them simple, easy directions and you valuable consumer data;**
- **The app offers a mailing list opt-in that provides the user with regular updates, menu**

items, specials, recipes, news, etc., and you with more valuable consumer data;

- **The app makes online ordering available, not just via calling in or text but an actual menu to order from, one click at a time, where users can select menu items, calculate the cost of order and choose pickup or delivery;**
- **A photo gallery featuring uploads from the restaurant itself as well as customers;**
- **An automatic coupon for downloading the app;**
- **Full "share-ability" on Facebook, Twitter and other social media channels;**
- **A tip calculator, "dictionary" for foreign foods like sushi, Thai, Mexican or Japanese, etc., and other handy, useful information;**
- **Feedback functionality for reviews, customer service, etc.**

All of these features encourage the customer to use the app more and more frequently, either to order online, purchase a gift card, make a reservation, find out about an event, etc. The more they use the feature, they more often they dip over into sharing materials on social media, providing feedback or generally just interacting with the place of business.

Whatever your business, be it a shoe store, nursery, nightclub or toy store, I'm sure you can see the implications for interactivity in such an app. These apps also give you the opportunity to update frequently, either with special offers, an upcoming event, new recipe, holiday greetings, birthday greetings, etc., upping the interactive nature of the relationship, one click at a time.

Apps aren't complicated, rather they are simple, attractive, functional and provide numerous benefits for both the business and the customer.

Parting Words: *Preparing for the Future*

Being more interactive isn't necessarily just about new technology, but new ideas. Many of our clients see mobile marketing as simply print or web advertising with a smaller screen. But that's also small thinking.

Instead, think of how those two words – mobile marketing – work together. Mobile is such a powerful element: this is what people are doing, right now, as we speak.

Think of unique and new ways to reach out to people, interact with them, and have them connect back.

CHAPTER 5:

ACTION STEPS

Here is a simple Action Step to make Chapter 4 more immediate:

Step 1: *Build a Mobile App to Engage Your Audience*

Mobile apps do not have to solve enterprise problems, they can be just as impactful by allowing your most loyal customers to view your menu, order products and services, set appointments, make reservations, integrate social connections and more.

Consumers enjoy using mobile apps over websites because of the speed, easy navigation and convenience of having the entire Internet in the palm of their hands – everywhere they go.

There are many do it yourself solutions available to accomplish this feat, such as ApporClick to take advantage of a 30 Day Free Trial to build you own mobile app go to mobile.simplyflex.com. I have also listed several other companies that offer mobile app builders to help you determine the best route to take in developing a mobile app for your business.

To see more examples of how mobile can be interactive, visit simplyflex.com/chapter5.

CHAPTER 6 - MOBILE IS LOYAL

Loyalty Programs and Your Business

Yesterday's home runs don't win today's games.

–Babe Ruth

Many of our clients balk at the time, energy and expense of creating a mobile application because they feel it's simply another extension of their offline, even online, marketing efforts. And, in the hands of some marketers, mobile apps are merely extensions of a website, blog or banner ad.

But in the right hands, and with the right strategy to complement it, a mobile application can increase both your brand recognition *and* loyalty in ways that far exceed your initial investment. This chapter details how:

Why Your Business Needs A Mobile Loyalty App

Many small business owners are missing out on the benefits of being able to reach out to their existing customer base by targeting them with initiatives that are guaranteed at increasing sales. That's because they lack a solid mobile loyalty app program that can help them leverage their existing customer base.

It is no secret to any business owner that existing customers constitute the most loyal client base that you can hope to have. Your existing customers are already familiar with your products and services and, therefore, are much more likely to be receptive to any promotions that are targeted at them.

Having a mobile loyalty app can help you target these existing customers, in a way that is both convenient and accessible to them, and helps you increase sales – and brand loyalty – in the process.

The Advantages of Having a Mobile Loyalty App

There are some very specific advantages to having a mobile loyalty app:

- A loyalty app can help you increase the number of visits to your mobile website.
- Loyalty apps are great for gathering information and creating profiles for your loyal, target customers.
- Such apps also allow you to track, monitor, measure and eventually influence consumer buying behaviors.
- Immediately alert consumers to new and unique discounts, rewards and special offers.
- Statistics show that as many as 35% of customers will go out of their way to a business that offers a loyalty program.

- A mobile app program can help bring in visitors more frequently.

The success of your loyalty program is likely to be stronger when you combine these rewards with timely user notifications that help target users through their smartphones.

More Loyalty = More Profits

Loyalty programs can also increase sales for you because they encourage spending. It isn't rocket science to assume that customers are more likely to buy something if they are offered gifts, offers or discounts for doing so, particularly from a company they've already purchased from.

While customers may not act on every special offer, discount, gift, coupon or reward, the continuity and convenience of being alerted to such offers will eventually move them to act, i.e. spend, with more and more frequency.

Loyalty programs encourage customers to spend more, which naturally leads to more spending, increasing sales for you. When customers are incentivized to spend more through rewards, gifts and free offers, they don't think of going anywhere else.

Offering Loyalty to Entice New Customers as Well

As the name would imply, a mobile loyalty app is generally targeted toward an existing customer base, but it can definitely help you generate new customers as well. Many customers will give up their old buying habits, and abandon their old retailers when they find some other businesses offering them much better deals through a loyalty app program.

In fact, according to statistics, about 40% of consumers admit to changing their spending habits to maximize loyalty rewards. In other words, your mobile loyalty program can incentivize customers to shift their spending toward your business to take advantage of the rewards that you offer. These rewards are not just a powerful inducement for your existing customers, but also a very strong lure for new customers.

Mobile loyalty can also help bring back old customers who may have shifted away to other businesses. There is nothing to bring back an old customer like a mobile loyalty program that rewards customers with gifts, discounts and other offers, when they spend more at your business.

Monitoring and Measuring: *The Foundations of Customer Loyalty*

What's more, loyalty apps provide ample opportunity for you to record, monitor and measure consumer spending behavior, not only through the app, your website or on-site. As any business owner knows, such information – freely given and easily accessed – is almost as valuable as the sale itself.

That's because a mobile loyalty app often comes designed with a customer activity tracking tool that leverages the power of analytics to gain more insight into customers. The more knowledge you have about customer spending and buying practices, and the manner in which they shop for products and services, the better positioned you are to be able to meet that demand for products and services.

A customer activity tracking program gives you access to information about your customers, especially about when and how they decide to make purchases, so you can design reward programs that work best for maximum results. This helps you plan your next marketing strategy more effectively.

Having a loyalty program is also an excellent way of increasing customer satisfaction, and making sure that current clients don't drift away to your competitors. In fact in a recent study, about 60% of customers reported that they were much more likely to continue shopping with a company that has a loyalty program in place.

Many marketers find that one of the major obstacles to businesses implementing a mobile loyalty program is the belief that having all these gifts and rewards in place will erode profits, and will not make any difference to a company's balance sheet.

That is a myth, and very often, the cost of having a rewards program is much lower than many business owners seem to believe. In fact, redemption rates are far lower than the 100% that many business owners fear. According to estimates, as many as one-third of all loyalty rewards are never redeemed.

Offering your customers a mobile loyalty app is long overdue, and many businesses have already realized that a loyalty program can be a strong sales generation tool.

Many customers now expect rewards and loyalty gifts when they shop at even small businesses, and base their loyalties on the kind of loyalty and reward programs these businesses offer. In a situation like this, can you really afford to miss out?

Customer loyalty programs are mainly designed with existing customers in mind, but these can also be a very effective tool to help attract new customers.

How Loyalty Programs Can Help Retain Customers

Customer loyalty programs have been around for eons, because businesses have always understood that it costs much less to retain existing customers than it does to go out and find a new customer. With smart phones and the emergence of mobile technology, it's much easier to retain customer loyalty. One of the best ways to build loyalty is by offering a customer loyalty program that incentivizes people to remain loyal to your business, and rewards them for doing so.

It's no secret that any customer is more likely to continue doing business with a company that actually rewards loyalty, and acknowledges people for staying loyal to the company. Nobody likes to be taken for granted, and a company that offers a loyalty program is much more likely to have a special place in the customer's heart. This also means that the customer is much more likely to think of your company, when he is in need of your products and services.

Having a customer loyalty program is one of the simplest ways to increase sales. The fact is that when we are promised rewards for doing something, we are much more likely to do that thing. That is exactly how it works with a customer loyalty program. When you offer or promise rewards to a customer for shopping with you, he's much more likely to do so because of the benefits attached.

Customer loyalty programs are mainly designed with existing customers in mind, but these can also be a very effective tool to help attract new customers. Freebies, gifts and offers that are part of a loyalty program can help attract customers who may be currently loyal to another company.

Ultimately, customers are loyal to the company that promises the maximum bang for their buck in terms of freebies, discounts, offers, deals and coupons. That is a basic sales fact, and customers will see no problem in moving on from their current favorite to a new company that offers them a much better deal. In fact, according to data, approximately 40% of consumers admit freely that they would consider changing their purchasing habits based on loyalty rewards.

When you have a customer loyalty program in place, you are much more likely to be frequently in touch with the customer. Studies show that customers are much more likely to visit a store that they have a loyalty program with, and visit the store much more often. To build on this fact, you can leverage the power of push notifications and maintain constant interaction with customers.

If you are launching a new brand or product or in the market, for instance, there is no better way to push it than with the help of your existing customers. These products can be added to your customer loyalty program by giving the product away as a freebie, or a contest prize, or gift. This increases product awareness for you at very little cost.

Overall, customer satisfaction levels continue to remain high when customers are enrolled in a loyalty program. In a recent study, 62% of customers admitted that they would continue doing business with a company with a loyalty program.

Customer loyalty programs increase customer interaction with your company, thereby giving you valuable data to base your analytics and customer tracking activities. You learn more about customer behavior as related to your product or service based on customer visits to your business, and gain access to other crucial bits of data that can help you plan out better strategies in the future.

Digital Coupons: *Mobile Incentives for Customer Loyalty*

According to new research, 92.5 million people in the United States redeemed digital coupons in 2012. Those are staggering statistics, and they are even more overwhelming when you begin to realize that the use of smart phones to redeem digital coupons has only existed for a few years.

When the recession hit in the late 2000s , folks were more eager than ever before to shop for discounts and deals. This was also the time that smart phone use increased in the United States, and many consumers found it very easy to access their coupons online.

Smartphones have changed the way that we do a number of things, from the way that the access the Internet, to the way we look for directions when we travel. In the same manner, smart phones have also changed the way that we redeem our coupons. There is no doubt that digital coupons are also here to stay. In fact, the popularity of these discount coupons is only likely to increase over the years ahead.

While the number of users who choose to redeem discount coupons via their desktop computers is still high, mobile users are pushing most of the growth in digital coupons. Some experts estimate that there will be a 4.6% growth in the use of discount coupons in 2013, and that the growth will remain fairly stable throughout the whole of next year.

One of the major advantages with mobile digital coupons is that they allow smartphone users to redeem their group coupons based on where they are at any point in time. A number of websites and businesses have seized upon the opportunity to provide discount coupons, and deals based on local customers and their locations at any given time, thereby increasing both their internal profits and customer loyalty at the same time.

Another interesting fact is that people who use discount coupons on smart phones, are not those who are trying to stretch their budget, but the affluent class. Earlier, discounts were in the form of paper coupons, and thrifty housewives would clip these coupons, intending to use them to save on groceries and household essentials. Now however, it isn't just about trimming the budget.

Modern customers who redeem digital coupons tend to be young, affluent, upwardly mobile, and interested in saving thousands of dollars not on groceries, but on international travel, premium luggage and other luxuries. In fact, according to statistics released in 2010, most discount digital coupon users came from households with incomes of greater than $70,000 a year.

That is not exactly a lower income household. In fact, households with an average income of approximately $100,000 or more were mainly responsible for pushing coupon growth in 2009. Many households with an average income of between $50,000 and 69,000 are also regular users of the mobile discount coupons.

It's not just that discount digital coupons are easy-to-use and so convenient, but also the fact that these coupons offer very good value for money. For instance, it's not unheard of for customers to get hundreds or even thousands of dollars off on an international holiday trip. Discounts up to 60% are fairly common.

These are not the "75-cents off" kind of coupons that your mother or grandmother used to cut out from newspapers. These are heavy discounts on products and services that do not exactly constitute necessities.

Another fact that separates digital coupons from their earlier versions is the age of the persons using the coupons. Earlier this month, new research released by RetailMeNot found that persons between 18 and 34 were much more likely to use digital coupons, than persons above the age of 35. In fact, consumers of this age were approximately 3 times more likely to use a mobile coupon. Not only are mobile coupons popular, but they're also very popular among a category of persons known for its propensity to shop and spend.

If you are a small business owner who wants to offer discount coupons, but does not know where to start, or does not believe that he has the finances to support the campaign, don't worry. Customers don't expect massive discounts with mobile or digital coupons, in order to redeem these. In fact, for many digital coupon customers, even a discount of just 25% off is considered a really good deal.

Parting Words: *Can Small Business Owners Really Afford to Ignore Mobile Apps?*

Cost-effectiveness, ease of use, potential for unlimited and efficient customer interaction and increased sales – these are just some of the reasons why a mobile app makes great economic sense for small businesses. In fact, the question should not be whether you as a small business owner can afford to ignore apps, but WHEN you plan to launch your app. If you need more reasons why you need a mobile app for your business, read on:

The fact is that as a small business, your business now needs mobile apps more than ever. Mobile apps are the mantra for modern advertising strategies, and reflect the winds of change blowing through the ways people access the Internet. Data indicates that most people now access the Internet using their smartphones and tablet computers, and if you want to reach out to these customers, you must make sure that your company is right there on their mobile screen.

For instance, you can use push notifications constantly to reach out to customers. Push notifications can be used to deliver customized messages to customers at specific times of the day.

Your mobile app can also be customized to make it easy for customers to buy products and services from your app. It can be frustrating for a customer who is interested in your product to leave your app and log into your website to place a product or pick up a phone to call up a customer representative. Your app must make it easy for customers to place orders on the app, smoothening the process of billing, payment and checkouts. Fortunately, many modern mobile apps now offer all of these features in an easy-to-use and safe format.

Mobile apps can also help you maintain constant customer interaction, helping you to increase the level of customer service that you offer. Customers can contact you instantly to offer recommendations, give valuable feedback or register a complaint. When customers find it easier to get in touch with you, they're much more likely to remain loyal customers.

Social media has changed the way people interact with each other, and your mobile app can make it even easier for your company to interact with customers. Mobile apps can allow people to share information and content on your Facebook pages, and upload pictures on your photo gallery. You can use your app to promote competitions and contests, and develop other ways for customers to interact with your company.

All of this helps promote brand loyalty, and strengthens your relationship with your customer. A loyal customer is much less likely to be lured by a competitor.

CHAPTER 6:
ACTION STEPS

Here is a simple Action Step to make Chapter 5 more immediate:

Step 1: *Build in a Loyalty Component*

Now that you have implemented a mobile app, make sure there is a loyalty component built in to make it more effective. For example, you can reward customers for checking in. This could be something as simple as a digital stamp card where customers show their mobile app that says "click stamp here" and the manager or store owner confirms the card with a code and your customer comes that much closer to receiving the reward.

Being that it is integrated into the mobile app as a strategy, you can send a push notification strategically on a Friday night and say something like, "Come visit us for an additional 10% if you participate in our mobile app rewards program".

Mobile loyalty needs to be integrated into all of your advertising because mobile makes it so easy to implement mobile coupons that drive foot traffic, and build your list very effectively.

CHAPTER 7 - MOBILE IS LOCAL

Level the Playing Field with Mobile Marketing

Winning is not a sometime thing; it's an all time thing. You don't win once in a while, you don't do things right once in a while, you do them right all the time. Winning is habit. Unfortunately, so is losing.

– Vince Lombardi

In just a couple of years – less than a single decade, in fact – mobile marketing has become one of the premier ways to attract new customers and retain existing ones. What was once believed to be a fad has instead firmly established itself as a must-implement strategy for any company, from a small business enterprise to a large retailer. According to ComScore, there are "over 100 million smart phone users" in the US.

Recently at the CTIA 2013 Mobile Marketplace conference, a number of executives voiced their opinions about how they believe mobile marketing technologies will continue to grow in the years ahead. One thing is clear, marketers are required now not only to adapt to smaller smartphone and tablet screens and a variety of programming languages and operating systems, but also to make these technologies much more simple to use, more convenient, and more consumer-centric.

Mobile marketing in the future is likely to offer much stronger consumer engagement, drawing consumers in different ways. Over the next few years, mobile marketers expect to see companies engaging in more real-time interaction with consumers using mobile devices, further solidifying consumer relationships.

There's also likely to be a lot of progress in the use of intelligence to determine how consumers use their mobile, and the use of this information to enhance convenience for users.

The use of mobile devices over the next few years is likely to grow from something that is currently restricted to marketing teams and corporations, to tools that the average consumer uses to gain access to basic services. In other words, we are well on our way towards democratization of technology which sees more people using mobile devices to gain access to the Internet, shop for products and services, just in the way that the average American now uses the Internet to do so.

Companies that want to stay ahead with mobile marketing over the next decade must stay relevant in the features that they offer consumers, and make websites and apps as simple to use as possible. Another growing trend in the future will be the use of consumer feedback to upgrade technology. Consumer feedback will be critical to the development and successful use of mobile technology for promotions in the years ahead.

What's more, companies will now have access to more local customers than ever before. That is, if they can recognize and harness the power of mobile marketing to woo their local customers into their virtual and brick and mortar stores:

Mobile Coupons Redeemed To Exceed $10 Billion This Year: *Is Your Local Business Cashing In?*

The days of cutting coupons from newspapers seem to be behind us. U.S. researchers say that the number of discount coupons that were redeemed by users using their mobiles and other devices, will cross $10 billion this year. That is an increase of more than 50% over 2012.

The research conducted by Juniper Research should clearly indicate to the marketing community and retailers that coupons for mobile devices are here to stay. The use of these coupons is very clearly on the rise, and statistics prove it.

More and more retailers now understand that the old way of doing business by offering coupons in magazines and newspapers is quite dead. If you're not offering customers your discounts and coupons on smart phones, tablets and other mobile devices, you're clearly missing out on targeting large volumes of local customers to your online ad or promotion.

As the folks at Juniper Research put it, mobile marketing allows customers not just a chance to visit your e-store, but also to make purchases, and pay for them using a smart phone. Brand loyalty is being built, and satisfaction is being gauged, helping these retailers retain old customers, and attract new customers.

Specifically for local consumers, coupons make it easier for locals to find out about, visit and purchase from your store, where they might not be engaged otherwise from more traditional local advertising.

Another interesting finding that emerged from the Juniper Research study is that mobile coupons have a much higher redemption rate, compared to traditional print coupons. The overall volumes of coupons were much lower in the case of mobile coupons, but the redemption rates were much higher at 10%, compared to traditional print media and PC coupons. In the case of these traditional coupons, the redemption rate was just 1% or less.

Part of the reason for this is that retailers can personalize mobile coupons, and target specific individuals, so that they are more appealing, and touch a chord with every individual user. This has particular implications for local consumers who may be enticed to visit based on more personalized coupon information.

The data proves the mobile marketing message – you don't just target customers. You target customers who are very likely to be attracted to your product or service, and actually buy it. That isn't the case with traditional media where you target a whole community of people, such as readers of a certain magazine or website, without really gauging their interest in your product, or affecting their ability or inclination to buy your product or service. Targeting of potential customers is higher, sharper and more precise with mobile marketing, particularly in the local arena.

Imagine designing, implementing and tracking a local coupon based on an annual or purely local event, such as a city park anniversary or founding father's birthday or local celebrity's achievement. Such immediate attention to your local market builds, and rewards, brand loyalty in ways that generic, across the board coupon marketing simply can't touch.

The Juniper Research also threw up some interesting new trends in mobile coupons. For instance, customers are much more likely to hold onto their mobile coupons, and store them, using them later and paying for them using their mobile wallets instead of using it immediately after issue. With mobile wallets so easily available, and so popular nowadays, it becomes even easier for consumers to store their mobile coupons, and use them later, using digital payment methods, without any hassle.

Simply put, mobile coupons are very easy to use, because a person doesn't have to store them anywhere, such as in a special binder, folder, holder or the pocket of their purse or wallet, where they are likely to get lost or expire from lack of use. The message for customers is this

– you can breathe easy, because you will never miss out on these coupons if you are connected to your favorite retailers on mobile.

The lesson for local retailers is this – if you want higher conversion rates, greater engagement with customers, stronger brand loyalty and enhanced customer experiences (and who doesn't), you have to switch to mobile ways of doing business.

The sooner, the better:

Locals Only: *Helping Local Business With Mobile Marketing*

What concerns many of my clients is cost. We have been led to believe that not only is designing an effective mobile marketing campaign expensive, but

In order to differentiate your local business from competitors, it's important to come up with pieces of content that actually show your expertise compared to your competitors.

running, maintaining, measuring and monitoring it will be as well.

Contrary to popular opinion, a mobile marketing campaign for a small business does not have to be prohibitively expensive. In fact, most marketing budgets for small retailers tend to be very limited, making it imperative that business owners go out of their comfort zone, and make use of simple, cost- efficient, yet effective strategies to reach out to consumers.

Fortunately, the Internet makes it very easy for businesses to market their products and services without investing heavily monetarily. If your business does not have a marketing budget that equals the GDP of a small country, don't fret. There is a lot that you can do with free marketing strategies online that are just as effective, and in some cases, even more effective than expensive promotional campaigns.

The main goal about having an Internet presence is to make your presence known. You have to be out there. That means creating content that is interesting, valuable to consumers, and can be shared with people. Write blog posts, or release press releases that increase your outreach online, and help you reach out to more numbers of consumers.

Press releases can get picked up by media outlets online which means great exposure for you. To make sure that your press releases are picked up, make sure that the releases are centered on a major event in your company, like the development of an app, or the launch of a mobile payment getaway that allows consumers to shop from their mobile devices.

Share your content as much as possible on social media. This is a simple strategy that is extremely cost-efficient, and very effective. Create a game that can be shared on Facebook to reach more consumers. Be selective in the number of posts that you share, and don't spam fans on your account.

Bring an experimental, innovative, and personal touch to the kind of content that you share on social media. For instance, it helps to post positive reviews of consumers who use your products and services. Consumers often very often rely on positive feedback from older customers of the company before they make a decision, and such reviews can definitely help. Rely on feedback from past consumers, and feature this on your Facebook page.

In order to differentiate your local business from competitors, it's important to come up with pieces of content that actually show your expertise compared to your competitors. For instance, you could develop a video that can be uploaded on YouTube, demonstrating how your business is much better than your competitors.

The key to making such videos successful is making them interesting to watch and share. It's also important to make sure that you demarcate yourself from your competitors, and outline your expertise as compared to other companies in your field, in order to impress consumers.

Everybody loves a contest, especially when it involves a free giveaway. Have contests online, and use your Facebook page to promote these. Make contests difficult enough people to get competitive and easy enough for people to actually participate. Gift people not just for winning the contest, but also participating.

Fortunately, none of the above strategies require heavy financial investment, but are proven, cost- efficient and effective ways of expanding your customer base.

Mobile Has Big Role to Play in Cross Device Advertising

Cross device advertising campaigns have taken off as companies decide to leverage smart phones, tablet computers and other devices to target local customers and get their message across platforms in a timely and efficient manner. This year especially, there has been great progress in expanding the scope of cross device advertising, as tablets continue to become more popular, and more consumers engage with a variety of devices according to their convenience.

Now more than ever, marketers across the country are looking at targeting not just customers who are on their smart phones, but also those who use their tablet computers at home, or move onto a laptop or desktop in the office. It's important to keep your message consistent, while also making sure that your local customers are targeted across all devices.

Right now, the technology is still in its nascent stages, but as technology becomes more advanced, and provide more information about a person's geolocation and other details, companies are likely to become more savvy about targeting consumers using mobile ad serving and tracking abilities. Already, companies are coming up with better and more engaging ad units that can target consumers efficiently across devices, and deliver messages while making sure that each message is uniquely presented across platforms.

The point is not to have the same template advertising flash across all devices. The aim is to capture attention and target customers across devices, delivering the same message in a different manner.

All consumers are likely to have mobile devices in their possession at any point in the day. Whether a person is at work, or at home, shopping or eating out, or at a sporting event, he's very likely to have a mobile device in his home. Computers and tablets are less likely to be used, and for the most part, data seems to indicate that tablets and laptop computers are still used mainly at home.

There are important implications in all this seemingly mundane data for marketers. Mobile marketers have now realized that the need to constantly target customers regardless of what device they are using at any particular point in time.

It is now becoming clear to marketers that cross device advertising that allows them to get their message across consistently, and literally 24/7, is a effective, and cost and time-efficient. No longer is it necessary to book the highest and priciest TV spots to target your local customers, as they wind down in front of the TV after long day. With cross device advertising, you can frequently stream your message across devices, and rest assured that your customer will be constantly bombarded with these messages, increasing campaign "stickiness".

Many people now use their smart phones for shopping. A person may start browsing for one item or product on a smart phone, and actually continue the buying process later. For instance, he may go home where he has more free time, and decide to access the Internet from his tablet computer.

Modern marketers need to be prepared for such changing browsing and buying behaviors by consumers, and update their advertising strategy accordingly. It is no longer enough to focus only on mobile technology, but also important to expand your advertising strategy across devices

It's also important to keep in mind that the marketing strategy must be expanded and customers need to be targeted, while keeping in mind user experience. Campaigns that simply blast the ad 24/7 can actually backfire if they do not translate into a customer- friendly user experience.

It is now also becoming easier for marketers to conduct their ad campaigns across devices. Recently, Google Enhanced Campaigns launched a new campaign that allows marketers to execute their paid search campaigns across devices including smart phones, tablets as well as desktops, using a single campaign.

Such campaigns are likely to become even more comprehensive and widespread in the days ahead. AOL Inc and Publicis Group also recently collaborated to create a new program called Publicis AOL Line, to execute advertising across mobile tablet, smart TV and desktop computers.

Such cross device campaigns are becoming very successful, which is why they're becoming so popular.

However, there are several challenges in implementing cross device campaign. For instance, there still remain issues like the appropriate tracking of the success of such campaigns. It's harder to develop appropriate metrics that are based on the unique capabilities, of a number of different devices like desktops, tablets and smart phones involved. There also need to be more unified reporting systems that help track the success of the campaign across devices.

Using tracking cookies on mobile devices has also proved challenging and this could affect optimization efforts. Optimization plays a major role in such cross device advertising. However, marketers are working on these issues and these hurdles are likely to be ironed out soon.

Use Social Media to Boost Your Local Revenues

Today's modern landscape is a highly competitive one, and companies that don't manage to stay one step ahead of competitors in their advertising and promotions, can quickly find themselves losing out. These days, it is the organizations that do not invest in marketing to their local customers via social media, and reaching out to fans and potential customers on Facebook that find themselves losing out on revenues.

Business owners must understand the value of social media, and invest in connecting with fans – both existing customers as well as potential new ones – on Facebook, Twitter, Foursquare and other social media pages. These pages can be used to share content that is relevant to customers. The point of having a Facebook page is not to spam customers with discounts, deals, exclusive offers, and all kinds of other promotional messages. The point is to share information that is very relevant to customer needs, and benefits them in some way. There are a number of ways that a local business owner can do this because of the very nature of the business and the services it provides.

Don't obsess over the numbers of fans, friends or followers that you have. It is far more important to have fans that engage with you.

Let's say your local business is a day spa. Well, you're in luck: many other services that serve spas have begun to integrate social media-tools into their packages. For instance, the Spa Booker software comes with a tool that aligns with Facebook, Twitter and Foursquare.

SpaBooker can create a promotional officer within the account, and this can be directly uploaded on the Facebook or Twitter page, complete with a link that allows a customer to redeem the coupon.

There are also gift certificate apps that can be used to reach out to customers on Facebook. There are other services being offered by ResortSuite, which allows e-mail promotions to be forwarded to customers via social media pages.

Whatever business you're in, it's important to set up a content calendar that will specifically define the kind of content that will be posted on your social media page every day. Have a mix of different types of content presented for customers to mix things up, and keep customers interested in your page. Avoid being spammy. That is the easiest way to get your fans to unlike your page. Your Facebook page should serve as an online community that consists of your local customers, as well as their friends. With referrals from your fans, you'll find your revenues increasing.

Know your target audience, and what they're interested in. Once you know what your audience is all about, it's easy for you to run promotions, giveaways, campaigns, contests, sweepstakes and other campaigns that will lure customers.

Don't obsess over the numbers of fans, friends or followers that you have. It is far more important to have fans that are actually active and liking and sharing your posts, than it is to have many followers who simply ignore your page.

See if you can establish partnerships with vendors and others in your industry, to help reach out to more potential customers. For instance, in our spa example, you could tie up with a local vendor of hair coloring products, and host a joint contest or promotion. You might also team up with a local charity and donate part of the proceeds, making another valuable local connection. This is the kind of marketing initiative that is likely to have benefits for both of you.

Make no mistake: content is the backbone of your social media campaign, and any content that you post is immediately linked to your firm. That means that there must be no errors in the kind of content that you post, and no embarrassing gaffes made by an intern that could quickly go viral. In other words, there must be oversight of the kind of content that you post.

Your social media campaign will reflect on your company, and therefore, the reigns of your social media presence should not be handed over to an inexperienced college intern who may share something completely inappropriate or worse, offensive. There are dozens of examples from around the country involving companies that have used their Facebook pages inappropriately, with major damage resulting to their reputation.

Understand the kind of content that customers will like, establish a content calendar, and hire experienced persons to post on your Facebook page regularly, and you will find, over a period of time, that your content marketing strategy is working to increase footfalls.

Parting Words

It's important that we begin to gradually demystify the conception that mobile marketing is only global. While the very nature of technology and the internet makes our local efforts national, even international, they can still be very much local.

By now, this chapter should make it patently clear that mobile marketing truly is local, and that one of the best ways to tap into your local market is to reach customers where they are. Which is, increasingly, on their smart phones, tablets and all the latest mobile devices.

CHAPTER 7:

ACTION STEPS

Here is a simple Action Step to make Chapter 8 more immediate:

Step 1: Go Local

Look for every opportunity you can to engage your audience with mobile. I've covered alot of details in this book. Although the Internet allows your business to scan the globe - mobile is the great equalizer and can engage your local market like never before.

1. Set specific goals. Your mobile marketing objectives should, of course be closely tied to your business's overall marketing plan. The most effective goals should include:

- Sell more of your product or service (mobile landing pages)

- Drive traffic to your bricks and mortar location or website (make sure it's mobile friendly)
- Promote special offers with mobile coupons (smartphone users are now customers)

2. Adapt your email messages. Flashy and colorful HTML-based emails may work well on a computer screen, but not so much on an iPhone display. No one wants to wait around for images to load — in fact, many users opt not to download them as a default setting.

To reach mobile users, focus on crafting the most concise messages possible. Every email must be snappy and contain a compelling call to action. The job of nearly every sentence in your email is to get the recipient to read the following sentence, Move your customers from sentence to sentence and you'll have a conversion faster.

3. As you go about your business in town, check to see which local companies offer mobile apps. Look for menus at restaurants or business signs, download them and study them. See which apps work, and which don't, or even which features of which app work – and which don't.

Cherry pick those features that you find really useful – a map, movie times, quick reviews, reservations, recipes, menu items, contact list, whatever it may be – and include them in a "wish" list for your own mobile app. Oftentimes "test driving" the apps of other companies helps you see what works in real life – and what doesn't – while letting them do the R & D for you.

IMPORTANT FACTS

Before You Go... 15 Mobile Commerce Facts That Every Small Business Should Know

Mobile commerce may seem like a cool extra channel for generating sales, until you learn more about it. Then you see that it's quickly becoming the main channel for generating sales online and connecting with customers. In fact, if trends continue, there may come a day when m-commerce is referred to more simply as just "commerce."

Okay, that's overselling it. But m-commerce is a big deal, and we've got four facts to back that claim up. Read on, but be warned – once you hear more, you're going to want in. The opportunities are just too darn tasty.

1. Mobile commerce will reach almost $40 billion by the end of the year

Wow. Is that for real? Yes. Yes it is.

Mobile commerce is a massive, booming market, and it's increasing all the time. If your business doesn't have a piece of it, you may be losing total market share as others arrive to the party ahead of you.

And it's a pretty big party.

2. Everyone has a mobile device

And not just that, a huge amount of people have smart phones (roughly half of adults in the U.S.). This means that you're not addressing a small slice of the market when you invest in m-commerce, you're addressing the majority of the market. Mobile usage rates across the world are expected to increase exponentially, with predictions of 2 billion smart phone users by 2015.

Again, these trends show that m-commerce is not an extra, supplemental channel for producing sales, but one that is fast-becoming the central channel in the digital marketplace. Of course, not all products and industries are impacted by this fact equally, but if you haven't checked out the data for your business recently, it's high time you did.

3. Some of the most aggressive shoppers online are people between the age of 19 and 34. These young customers also form the bulk of all mobile searches.

4. 45% of users in this age category admit that they use mobile search features every day. That fact alone is going to have a tremendous impact on your business.

5. If you offer "store mode," customer engagement increases fivefold.

Yep, engagement is five times higher if your mobile strategy has store mode. Even when people visit your location in person, they're still visiting you on mobile. You just can't get away from it. It's everywhere. Literally. But there's good news:

It's a huge opportunity!

If your customers love mobile, then it pays for you to love mobile as well. And facts like this show just how much it pays. Your mobile dollar can go very far indeed.

6. 50% of American smartphone owners had admitted using mobile apps as part of their shopping experience.

7. Some shoppers only use mobile devices

This is the most compelling fact, in our opinion. Some shoppers can't be reached on desktop devices anymore – they only connect via mobile. If you haven't gone mobile, these customers are simply on another planet from you.

8. Smartphone users also admit they are likely to increase the amount of Internet access via smart phone in the future.

That means more numbers of people online, searching for products and services, and armed with the tools to purchase your product, if they like it.

9. There were approximately 150 million searches conducted on mobile devices in 2012. More than 50% had local intent.

10. 94% of smart phone users look for local data on their phone.

11. Mobile searches are expected to touch 86 billion in 2015, far surpassing the number of searches conducted on desktop computers.

12. Approximately 40% of all searches on mobile devices occur between the hours of 6PM and 12PM.

If you are new to investing in mobile technology, you must know that location and timing is key in reaching out to consumers.

More than three quarters of all mobile searches occur at home or at work, even surpassing your desktop computers.

13. Approximately 2 out 3 of all mobile searches that are conducted in stores are conducted to help the person make a decision to buy a product.

14. 70% of all searches on mobile devices lead to actions that lead to some type of conversion.

15. Approximately 40% of users would move away from a mobile site, if the site that they visited is not mobile-friendly.

Parting Words

I hope by now you understand that not only is mobile marketing the wave of the future, but that it's already here, live and in the present. I hope this book has helped you realize both the impact, and the opportunity, of mobile marketing on your own business and potential profit.

In fact, I would love to hear how you're progressing in your inclusion and expansion of mobile in your own business. I'd love to hear from you, please reach out to me on Twitter @billheneghan, email me at info@simplyflex.com or visit my blog at www.simplyflex.com to discover up to the minute research, facts, trends, statistics, information, suggestions, tips, strategies and applications of mobile marketing now and in the future.

ABOUT THE AUTHOR:

BILL HENEGHAN

Bill Heneghan is an entrepreneur, investor and marketing technology guru. Bill has over 16 years of experience involved in entrepreneurial ventures and start-up companies. Mr. Heneghan is an expert in identifying new market segments, innovative product offerings, and strategic initiatives related to market share and profitability. He has a passion for free enterprise and small businesses.

He has developed several successful companies and in 2007 was instrumental in revamping an eye care company that was producing $300,000 in revenue to over $30 million in 5 years, by converting a wholesale business model into a direct marketing powerhouse. As a former professional athlete he leads with a special focus on excellence, accountability, exceptional client experiences, and untraditional marketing. His passion is spending time with his family, helping others succeed in business and, of course, a little golf.

www.ingramcontent.com/pod-product-compliance
Lightning Source LLC
Chambersburg PA
CBHW050509210326
41521CB00011B/2380

* 9 7 8 0 6 9 2 2 4 7 0 5 1 *